Small Ships in Theater Security Cooperation

Robert W. Button
Irv Blickstein
Laurence Smallman
David Newton
Michele A. Poole
Michael Nixon

Prepared for the United States Navy

NATIONAL DEFENSE RESEARCH INSTITUTE

The research described in this report was prepared for the United States Navy. The research was conducted in the RAND National Defense Research Institute, a federally funded research and development center sponsored by the Office of the Secretary of Defense, the Joint Staff, the Unified Combatant Commands, the Department of the Navy, the Marine Corps, the defense agencies, and the defense Intelligence Community under Contract W74V8H-06-C-0002.

Library of Congress Cataloging-in-Publication Data

Small ships in the war on terror / Robert W. Button ... [et al.].
    p. cm.
  Includes bibliographical references.
  ISBN 978-0-8330-4400-6 (pbk.)
    1. Gunboats. 2. War on Terrorism, 2001–3. United States—Military policy.
  I. Button, Robert.

V880.S53 2008
623.825'80973—dc22

                                                            2007051919

Top Photo: PC-1 Cyclone. Courtesy of the U.S. Navy.
Bottom Photo: Siyu Primary School dedication ceremony speech is given by
Capt. Larry Flint, USN, on behalf of Rear Admiral Richard Hunt,
Combined Joint Task Force-Horn of Africa, CJTF-HOA.
Courtesy of the U.S. Navy.
Photo by Chief Mass Communication Specialist Paul Del Signore

The RAND Corporation is a nonprofit research organization providing objective analysis and effective solutions that address the challenges facing the public and private sectors around the world. RAND's publications do not necessarily reflect the opinions of its research clients and sponsors.

RAND® is a registered trademark.

Published 2008 by the RAND Corporation
1776 Main Street, P.O. Box 2138, Santa Monica, CA 90407-2138
1200 South Hayes Street, Arlington, VA 22202-5050
4570 Fifth Avenue, Suite 600, Pittsburgh, PA 15213-2665
RAND URL: http://www.rand.org/
To order RAND documents or to obtain additional information, contact
Distribution Services: Telephone: (310) 451-7002;
Fax: (310) 451-6915; Email: order@rand.org

Preface

The U.S. Navy is considering the potential capability of small ships to contribute to theater security cooperation (TSC) activities in the War on Terror and other identified tasks. The Assessment Division of the Office of the Chief of Naval Operations (OPNAV N81) asked the RAND Corporation to establish characteristics appropriate for small ships in this context.

This report presents the results of the research performed by RAND for N81. During their study, the authors developed a concept of operations for a small ship to be used in TSC with a set of additional tasks required for TSC and associated missions, identified characteristics of ships used by other nations in TSC, and listed commercially available alternatives.

This report should be of interest to the Department of the Navy, the U.S. Coast Guard, the Joint Commands, and Congress.

This research was conducted within the Acquisition and Technology Policy Center of the RAND National Defense Research Institute, a federally funded research and development center sponsored by the Office of the Secretary of Defense, the Joint Staff, the Unified Combatant Commands, the Department of the Navy, the defense agencies, and the defense Intelligence Community.

For more information on RAND's Acquisition and Technology Policy Center, contact the Director, Philip Antón. He can be reached by email at atpc-director@rand.org; by phone at 310-393-0411, extension 7798; or by mail at the RAND Corporation, 1776 Main Street, Santa Monica, California 90407-2138. More information about RAND is available at www.rand.org.

Contents

Figures

Tables

Summary

The U.S. Navy is considering the potential capability of small ships to contribute to TSC activities in the War on Terror and other identified tasks. The RAND Corporation's objective was to establish characteristics appropriate for small ships in this context. RAND was directed to focus its analysis on TSC operations in sub-Saharan Africa, with 21 countries of interest to the U.S. Navy identified.

In addressing this problem, RAND analysts found several questions that must first be answered, including the following:

- What is a suitable concept of operations for a small ship to be used in TSC? What other ship missions would relate to TSC missions?
- What additional missions should a small ship be prepared to conduct in order to obtain cooperation from potential partner nations?
- What tasks are required by the above missions? What capabilities are required to conduct these tasks? What ship characteristics are needed to provide these capabilities?
- What is the nature of the U.S. Navy's interest in foreign navies, and how would a small ship interact with these navies?
- What are the environmental constraints on ship characteristics? Do sea state considerations in regions of interest put a lower bound on vessel size? How do sea state considerations affect the concept of employment?

- What types of ships do other countries use in their TSC programs, and what candidate ships are available commercially? What can be said about cost?

Three general classes of vessels suitable for use as a small ship in TSC are identified in this report. Differing primarily in size and level of support required, they are the

- **Nearshore patrol vessel.** These vessels displace fewer than 100 tons. They require logistic and operational support, including the following: hotel services; refueling, rearmament, and re-supply; additional small vessel rotational crews; maintenance facilities and support; feed of situational awareness;[1] and provision of additional command, control, communications, computers, and intelligence (C4I) support.[2] *A specialized mothership would be required for these vessels.*
- **Coastal patrol vessel.** These vessels displace 300–700 tons. They require some logistic and operational support, including the following: regular refueling and re-supply; some situational awareness; and tailored C4I support. *These vessels would benefit from a mothership of opportunity.*
- **Offshore patrol vessel.** These vessels displace approximately 1,500 tons. They would benefit from some logistic and operational support, including the following: occasional refueling and re-supply; some situational awareness; and tailored C4I support. *These vessels require no mothership.*

These ship classes, their requirements based upon a task evaluation, and evaluations of the ship classes were arrived at using a "strategies-

---

[1] A nearshore patrol vessel, particularly one envisaged as a cheap, simple solution suitable for use by partner nations, would not have the systems or personnel needed to compile a sophisticated operational picture. The support vessel (or shore facility) would need to do this.

[2] Again, a very small vessel will not have adequate personnel to undertake wide area command and control or sufficient space to accommodate enhanced communications and intelligence equipment.

to-tasks" methodology. The strategies-to-tasks methodology identifies the tasks needed to achieve specific military objectives, starting with the National Military Strategy and cascading down through national military objectives, campaign objectives, operational objectives, operational tasks, functions, and, finally, operational systems.[3] This assessment of ship characteristics stops short of defining ship systems.

The nearshore patrol vessel is the smallest and least-expensive vessel with greatest access to shallow waters and minor ports. The low unit procurement cost would be offset for the U.S. Navy, however, by the need for a dedicated, specialized mothership to support this vessel. The nearshore patrol vessel has the worst habitability, would be the least survivable in rough seas or because of enemy action, and would be the least-capable vessel. Finally, while it is an attractive entry-level vessel to some nations, potential partner nations are now buying larger vessels.

The coastal patrol vessel is also a small vessel with good access to shallow waters and minor ports. Increased size gives this vessel numerous operational advantages over the nearshore vessel, including better survivability, greater endurance, and improved habitability. Coastal patrol vessels would not need a specialized mothership and could be supported by a suitable vessel of opportunity, although such a ship might not provide as much support as a specialized mothership. The larger size of the coastal patrol vessel would enable it to work more comfortably with the relatively larger vessels (over 1,000 tons) now being purchased by potential partner nations.

The offshore patrol vessel is the largest of the vessels considered in this study. It would have the greatest independence of operation and is the most capable and versatile vessel, able to undertake long patrols. In size, it is the most comparable to the larger vessels being purchased by potential partner nations. With these advantages comes increased cost, however. A fixed budget would allow fewer to be purchased, potentially reducing regional presence.

---

[3] See David E. Thaler, *Strategies to Tasks: A Framework for Linking Means and Ends*, Santa Monica, Calif.: RAND Corporation, MR-300-AF, 1993, for a more thorough treatment of the methodology.

Study Team Observations

In this quick survey report on small ships, the RAND team did not draw definitive conclusions; rather, we present several observations for the U.S. Navy to consider in a more definitive study of the small ships phenomenon and their employment in the War on Terror. First, the U.S. Navy needs to give more consideration to the constabulary needs of potential partner nations in order to gain the increased access it needs to undertake TSC (and, potentially, operations in support of the War on Terror).

Second, whatever small vessel is chosen, success in TSC will greatly depend on the qualities of the crew. Small ship personnel will need to be specialized (with language training, for example) and they will require stability in their assignments to assure adequate time in theater and to prevent untimely personnel rotations. In this way, they will stand a better chance of building long-term relationships with the navies of partner nations. We recommend that the crews of these vessels be specially selected, with skills akin to those of special operations forces teams, and that they be given specific training to improve their abilities in constabulary and TSC tasks.

Third, while specialized motherships offer advantages in terms of operational suitability, they also have the disadvantage of needing to be procured through a formal acquisition process that could delay the implementation of the small vessel concept. In addition, such vessels might have limited utility in wider combat operations and would represent significant additional cost to the U.S. Navy.

Fourth, there is value in quantity. Having many of these small ships would be more beneficial to the U.S. Navy's concept of a vessel for the War on Terror than fewer. A squadron of five ships, for example, could support each other and provide the necessary intelligence and command, control, communications, computers, intelligence, surveillance, and reconnaissance (C4ISR) meshing that a larger ship alone could not.

Fifth, most countries of interest are gravitating toward larger rather than smaller ships. If the U.S. Navy wants to operate with those navies in the future, a comparable ship should be considered. If the

U.S. Navy wishes to sell such a ship through the U.S. Department of State's Foreign Military Sales program, the larger ship again seems to be of more interest to potential partner nations (based on their recent ship procurements).

Sixth, while motherships (i.e., Navy combatants especially configured to support one or more small ships) offer advantages, they are costly and potentially vulnerable because they probably could not access the same ports as the ships they are designed to support.

Finally, the study team does not recommend the nearshore patrol vessel. Considerations of fuel, stores, and crew fatigue give it the least endurance of the three potential solutions. Crew fatigue will be exacerbated by the small crew size (approximately ten persons), which will also require steady crew rotation and the need for multiple crews in theater. It has the worst habitability, especially in difficult sea states, and is least survivable in terms of both seakeeping and vulnerability to small arms fire. It was seen as least capable of performing the overall constabulary and TSC missions, and would be most dependent on a dedicated mothership for capability.

This study is a preliminary analysis of new or changing missions that the U.S. Navy may face as it attempts to partner with maritime nations beyond those with which it has enjoyed longstanding relationships. It was not possible in this small study to be as definitive in our research and analysis as we might desire. Additionally, the U.S. Navy may wish to test and develop our concept of employment, looking at issues such as how to get the small vessels into a theater and support them once deployed. Other key steps might include determining the force structure of the mothership (number and type) and the small vessels. Finally, small vessels may be able to contribute to missions beyond those of constabulary and TSC.

Next Steps for the U.S. Navy

We suggest that the U.S. Navy validate and further develop a concept of employment for the small ships. The concept of employment should consider:

- how the small ship reaches the theater of interest
- how the small ship is supported in theater, including
  - the potential use of contractor logistic support
  - new manning options, including longer tours for crew
  - the concept of the mothership and its use, including cost, basing rights, load lists, etc.
  - the possibility of partner nation support, including the potential advantage of working in-country with a host nation and the potential disadvantage to force protection
  - issues of force structure, including the merits of squadrons for combined operations and the issue of support if the ships are not homeported or based in a host nation.

Finally, the potential roles for a small ship outside the TSC world should be examined.

Epilogue

Shortly after this analysis was briefed out to the U.S. Navy, the service was tasked to examine the PC-1 Cyclone Class as a small ship for use in TSC.[4] For use in TSC, the PC-1 was to be given an updated propulsion system and improved command and control for greater connectivity. A non-stabilized 25-mm gun was to be replaced by a stabilized 25-mm gun. The PC-1 displaces 331 tons, placing it toward the low end of the notional coastal patrol vessel band (300–700 tons).

At the U.S. Navy's request, the RAND study team conducted a short follow-on study of the PC-1 using data on these updates and improvements. We found that the PC-1 would be somewhat less capable than the notional coastal patrol vessel used in this study, but that mothership support would render it fully capable. This result was accepted for use by the U.S. Navy.

---

[4] The U.S. Navy transferred most ships of the PC-1 Cyclone Class to the U.S. Coast Guard in 2000. The U.S. Coast Gaurd then redesignated them the PC-179 Class.

Acknowledgments

This study was greatly aided by support from Stephen Farley, Robin Beall, and CDR Brian Hoyt of N81, the U.S. Navy's Assessment Division of OPNAV. Hans Pung was untiring in his research support for this project. Finally, CDR Brad Williamson, former captain of a Navy patrol boat, was especially helpful in our understanding of small boat operations around the Horn of Africa. We also recognize the efforts of our reviewers, Robert Murphy and Jennifer Moroney, who improved this report.

As always, any errors are the sole responsibility of the authors.

Abbreviations

C4I	command, control, communications, computers, and intelligence
C4ISR	command, control, communications, computers, intelligence, surveillance, and reconnaissance
DoD	Department of Defense
EEZ	exclusive economic zone
FNMOC	Fleet Numerical Meteorology and Oceanography Center
IMO	International Maritime Organization
NOC	Naval Operations Concept
OPNAV	Office of the Chief of Naval Operations
OPV	offshore patrol vessel
SOF	special operations forces
TSC	theater security cooperation
UNCLOS	United Nations Convention on the Law of the Sea
USCG	United States Coast Guard
WMD	weapons of mass destruction

CHAPTER ONE

Introduction and Objectives

Introduction

The United States entered the War on Terror following the events of September 11, 2001. The National Security Strategy, the National Defense Strategy, and the National Military Strategy have since been altered to accommodate the additional requirements and new priorities needed to win this war. With regard to the U.S. Navy, the Chief of Naval Operations established the following Navy missions in support of the War on Terror:

- Deny terrorists the use of the maritime environment.
- Enable partner nations to counter terrorists in the maritime environment.
- Deny the use of the maritime environment for the proliferation of weapons of mass destruction (WMD).
- Exploit the maritime environment in order to defeat terrorists and their organizations.
- In coordination with other government agencies and partner nations, counter state and non-state support for terrorism.
- Employ naval capabilities to counter ideological support to terrorism.

These missions will require the U.S. Navy to work with partner nations. In working with foreign nations, the U.S. Navy will, by definition, engage in security cooperation operations and associated activities.[1] The open nature of the maritime environment, which assures the U.S. Navy's access to the world's oceans, gives terrorists and their supporters wide access to it as well. International agreements limit the U.S. Navy's ability to operate in waters that belong to other nations, further requiring cooperation with those nations in the pursuit of terrorists or to counter their operations. The U.S. Navy could be well positioned to encourage and enable partner nations to exploit the maritime environment to pursue objectives to defeat terrorism. These missions are encapsulated in the concept of the "1,000-ship Navy" articulated by ADM Mike Mullen.

In 2004, ADM Thomas Fargo, then-Commander, U.S. Pacific Command, stated that theater security cooperation (TSC) "is the vehicle through which we extend U.S. influence, develop access, and promote competence among potential coalition partners. These activities directly support the War on Terror and enhance readiness for contingency actions against emerging threats."[2] Since his statement, the U.S. Navy codified security cooperation in the 2006 Naval Operations Concept (NOC), which states,

> While forward deployed, our Sailors and Marines will be critical members of the joint and inter-agency team that interacts with an expanding set of international partners to build defense relationships, develop friendly capabilities for self-defense and multinational operations, promote cultural awareness and regional understanding, and enhance strategic access. Always conducted with the utmost respect for individual national sovereignty, these

---

[1] The U.S. Department of Defense (DoD) defines Security Cooperation as "All Department of Defense interactions with foreign defense establishments to build defense relationships that promote specific US security interests, develop allied and friendly military capabilities for self-defense and multinational operations, and provide US forces with peacetime and contingency access to a host nation." DoD Security Guidance adds the objective of improving information sharing and intelligence sharing to help harmonize views on security challenges. Joint Publication 1-02, *Department of Defense Dictionary of Military and Associated Terms*, Washington, D.C.: Joint Staff, April 12, 2001, as amended through October 17, 2007.

[2] U.S. House of Representatives, *ADM Thomas B. Fargo, U.S. Navy, Commander, U.S. Pacific Command: Testimony Before the House Armed Services Committee*, March 31, 2004.

cooperative activities will include assisting host nation govern-
ments in freeing and/or protecting their societies from subver-
sion, lawlessness, and insurgency.[3]

The NOC later adds,

On a routine basis, our Sailors and Marines will likely operate together from a variety of ships, aircraft and small boats to
conduct maritime security operations in sea lanes, coastal areas,
and rivers in order to protect the world's waterways from terror-
ist activity, piracy, drug smuggling, and environmental threats.
These activities will likely involve security cooperation with the
forces of nations with similar goals.[4]

Study Objectives

The U.S. Navy is considering the potential capability of small ships
to contribute to TSC activities in the War on Terror and other identified tasks. The RAND Corporation was asked to establish characteristics appropriate for small ships in this context. RAND was directed
to focus its analysis on TSC operations in sub-Saharan Africa, with 21
countries of interest to the U.S. Navy identified.

In addressing this problem, RAND analysts found several questions that must first be answered:

- What is a suitable concept of operations for a small ship to be
used in TSC? What other ship missions would relate to TSC
missions?
- What additional missions should a small ship be prepared to
conduct in order to obtain cooperation from potential partner
nations?

---

[3]   Department of the Navy, *Naval Operations Concept 2006*.

[4]   Department of the Navy, *Naval Operations Concept 2006*.

- What tasks are required by the above missions? What capabilities are required to conduct these tasks? What ship characteristics are needed to provide these capabilities?
- What is the nature of navies of interest, and how would a small ship interact with them?
- What are the environmental constraints on ship characteristics?
- What sorts of ships do other countries use in their TSC programs, and what candidate ships are available commercially? What can be said about cost?

This report describes the results of our study of the characteristics of such a small ship. It further describes the strategies-to-tasks methodology used in the study. The report concludes with findings, observations, and recommendations to the U.S. Navy for next steps in moving forward. The conclusions of this brief study are not definitive; rather, we provide observations for the U.S. Navy to consider in more definitive studies of the use of small ships in TSC operations.

Organization of This Report

The organization of this report reflects the study's underlying strategies-to-tasks methodology. Chapter Two introduces an appropriate concept of employment to frame required missions and tasks for the small ship. Chapter Three translates required tasks to capabilities needed for the small ship. Chapter Four derives possible combinations of ship characteristics implied by required capabilities. This entails a survey of those foreign navies with which the small ship may interact. It also entails consideration of the physical environment in which the small ship may operate. Chapter Five presents the results of a global survey of ships that have the desired characteristics to better understand what other characteristics the small ship would have. Candidate vessels and their cost implications are then described. Chapter Six draws together findings, observations, and recommendations from the study. Chapter Seven, an epilogue, describes a short follow-on project. Appendix A provides material from our environmental analysis, which, by direction, focuses on Africa. This appendix expands the environmental analysis globally to indicate small ship operability worldwide. Appendix B provides, for completeness, an expanded task list for a small vessel conducting TSC operations.

CHAPTER TWO

Linking Missions and Tasks

Background

Our study used a strategies-to-tasks approach to develop the character-istics required of a small vessel conducting TSC operations. The ben-efit of this approach is that it links high-level national objectives to the capabilities of specific systems through the development of task lists.[1] In preparation for this project, OPNAV N816 independently produced a task list for TSC using an established breakdown of mission areas and associated tasks. These two task lists are reconciled under the strategies-to-tasks methodology to produce a capability set for both task lists.

The N816-Generated Task List

N816's task list included many tasks that a small vessel will never undertake, such as providing area air defense. Because these tasks will be the responsibility of other, more militarily capable units, or will not be associated with small vessel operations, we eliminated them from consideration. The resulting list, shown in Table 2.1, was resolved alongside our own for comparison and as a crosscheck for complete-ness.[2] It was accepted for this study.

---

[1] David E. Thaler, *Strategies to Tasks: A Framework for Linking Means and Ends*, Santa Monica, Calif: RAND Corporation, MR-300-AF, 1993.

[2] The reconciliation is presented in Chapter Five.

Table 2.1
Selected TSC Missions from an N816-Provided Task List

Navy Mission	Capability
Amphibious warfare	Plan/direct use of mobile training teams to provide instruction to non-U.S. units
	Conduct use of mobile training teams to provide instruction to non-U.S. units
Information operations	Conduct psychological operations broadcast
	Plan/conduct operations security
	Plan/conduct psychological operations in support of decrease in transnational terrorist networks activity
	Plan/conduct psychological operations in support of decrease in smuggling operations
Mission of state	Conduct force/unit tour for foreign dignitaries
	Conduct systems/weapons demonstrations for foreign dignitaries
	Conduct foreign port calls
	Conduct receptions for foreign dignitaries during port calls
	Provide volunteers for small project assistance during port calls
	Participate in military exercises with allied nations
	Participate in military exercises with non-allied nations
	Conduct maritime interception operations and/or visit, board, search, and seizure operations with naval/combined/joint forces
	Provide training and advice on how to reduce vulnerability to terrorism and other threats, particularly in the maritime environment
Non-combat operations	Provide explosive ordance disposal assistance
	Conduct operations with other national governments

The RAND-Generated Task List

Joint Publication 3-57.1, *Joint Doctrine for Civil Affairs*, categorizes engagement activities under TSC. Briefly, in the context of this analysis, these categories are

- **Operational activities.** These activities include routine and continuing operations, not crisis response or episodic activities of an emergent operational nature. Examples could include missions using forces present overseas (such as peace operations, foreign humanitarian assistance, sanctions enforcement, and counterdrug operations).
- **Security assistance.** Security assistance activities include international military education and training. Security assistance is a significant aspect of the combatant commander's theater strategy.
- **Combined exercises.** This category highlights the nature, scope, and frequency of peacetime exercises designed to support theater, regional, and country objectives. Exercises could include civil-military operations (such as road building, school construction, and medical, dental, and veterinarian civic action projects).
- **Combined training.** This category includes scheduled unit and individual training activities with forces of other nations. Joint Combined Exchange Training is a special category of combined training that involves U.S. special operations forces (SOF) training with the armed/security forces of a friendly foreign country. Joint Combined Exchange Training is designed to give SOF the opportunity to accomplish mission-essential task list training. Improved interoperability with foreign forces participating in the exercise is an additional benefit of a Joint Combined Exchange Training activity.
- **Combined education.** This category includes activities involving the education of foreign defense personnel by U.S. institutions and programs in the continental United States and overseas.
- **Military contacts.** This category includes senior defense official and senior officer visits, counterparts visits, ship port visits, participation in defense shows and demonstrations, bilateral and

multilateral staff talks, defense cooperation working groups, military-technical working groups, regional conferences, State Partnerships for Peace, and personnel and unit exchange programs.

- **Humanitarian and civic assistance.** This category consists primarily of humanitarian and civic assistance provided in conjunction with military operations and exercises, assistance in the form of transportation of humanitarian relief, and provision of excess non-lethal supplies for humanitarian and civic assistance purposes. Other forms of assistance, such as demining training, also may be applicable to this category.
- **Other engagements.** This category consists of engagement activities conducted by the combatant commander or executive agent that do not properly belong in one of the previous categories. Examples include activities planned as part of the implementation of the provisions of arms control treaties and other related obligations.[3]

These categories are far too broad for use in determining the physical attributes needed of a small ship to conduct TSC missions. Thus, we elected to use a strategies-to-tasks methodology. The lexicon of TSC planning is not used here for several reasons. Use of TSC terminology would require an unacceptable restatement of the U.S. Navy's task list for the small ship. Further, the terminology of TSC is not compatible with the strategies-to-tasks framework used in this analysis to determine attributes of a small ship for TSC. Recasting this framework in the lexicon of TSC would diminish clarity.

Study analysts developed a task list for this study by examining national security doctrine (including the National Security Strategy, the National Defense Strategy, and the 2005 Quadrennial Defense Review) and joint and single service documents. This analysis yielded the following primary military objectives within the strategies-to-tasks framework:

---

[3] Joint Publication 3-57.1, *Joint Doctrine for Civil Affairs*, Washington, D.C.: Joint Staff, April 14, 2003, pp. vii–2.

- strengthen alliances and partnerships
- increase partner nation capacity to defend both itself and its mutual interests
- counter ideological support for terrorism
- build stronger security ties with Muslim countries
- change misperceptions of the United States and the West
- identify common interests and enhance our partners' ability to operate with U.S. forces through combined training and information sharing
- provide long-term (multiyear) presence in theater.

The interests of potential partner nations are not reflected here. However, those interests must be considered in order to achieve international cooperation with the U.S. Navy in TSC. The small ship and the services it provides must appeal to potential partner nations; at minimum, a partner nation will want to benefit from the presence of a U.S. Navy small vessel. The nation and its maritime forces will want to be able to relate to the capabilities of such a vessel and understand, even witness, how it can meet national needs. Ideally, the vessel will be suitable for a partner nation's own use through purchase or international assistance arrangements.

Use of the Sea

Various theoretical constructs can be used to describe the maritime needs of nations. For this study, we adopted a framework known as the "Use of the Sea"[4] and drew on enduring ideas of maritime function that can describe the interests of all maritime nations. These nations vary in "power," as can be seen in Table 2.2. Our framework enables reasonable comparison of the needs of the U.S. Navy, ranked first, with those of the potential partner nations, who rank last.

Constabulary and token navies function at levels that may be difficult for Americans to appreciate. At one extreme, Angola's navy has not had a seaworthy vessel since the year 2000. The navies of Palau,

---

[4]   A more complete discussion of the roles of navies and Use of the Sea can be found in Eric Grove, *The Future of Sea Power*, Annapolis, Md.: Naval Institute Press, 1990.

Table 2.2
A Ranking of Navies

Rank	Examples
1. Major global force projection—complete. This navy is capable of carrying out all the military roles of naval forces on a global scale. It possess the full range of carrier and amphibious capabilities, sea control forces, and nuclear attack and ballistic missile submarines in sufficient quantity to independently undertake major operations.	United States
2. Major global force projection—partial. These navies possess most if not all of the force projection capabilities of a "complete" global navy, but only in sufficient numbers to undertake one major "out of area" operation.	Soviet Russia (1990)
3. Medium global force projection. These navies may not possess the full range of capabilities, but have credible capacity in some of them and consistently demonstrate a determination to exercise them at some distance from home waters, in cooperation with other force projection navies.	United Kingdom, France
4. Medium regional force projection. These navies possess the ability to project force into the adjoining ocean basin. While they may have the capacity to operate farther afield, they do not do so on a regular basis.	China, India, Japan
5. Adjacent force projection. These navies have some ability to project force well offshore, but are not capable of carrying out high-level naval operations over oceanic distances.	Israel
6. Offshore territorial defense. These navies have relatively high levels of capability in defensive (and constabulary) operations up to about 200 miles from their shores, having the sustainability offered by frigate or large corvette vessels or a capable submarine force.	Egypt
7. Inshore territorial defense. These navies have primarily inshore territorial defense capabilities, making them capable of coastal combat and constabulary duties. This implies a force comprising missile-armed fast-attack craft, short-range aviation, and a limited submarine force.	Oman, Singapore
8. Constabulary. These significant fleets are not intended to fight, but to serve a purely constabulary role.	Sri Lanka, Jamaica, Mexico, Ireland
9. Token. These navies have some minimal capability, but this often consists of little more than a formal organizational structure and a few coastal craft. These states, the world's smallest and weakest, cannot aspire to anything but the most limited constabulary functions.	Cambodia, Angola, Kiribati, Madagascar, Palau, Tuvalu, Nigeria, Western Samoa

Tuvalu, and Western Samoa each consist of a single small patrol boat. At the other extreme, some navies of interest to the U.S. Navy operate modern combatants displacing thousands of tons.[5] These extreme cases bookend the navies of interest. Regardless of their capabilities, navies have the following three roles under the Use of the Sea concept:[6]

- **Military role.** These activities are usually associated with war or combat operations. They draw on the ability of a navy to exercise sea control and sea denial, and lead to power projection ashore.
- **Diplomatic role.** These activities support national policy, generally beyond territorial and economic boundaries. They increasingly include support to international policies adopted by recognized institutions such as the United Nations. They draw on the presence of naval forces and latent naval capabilities.
- **Constabulary role.** These activities enforce laws and treaties, generally within national boundaries or areas of national influence. The constabulary role includes economic enforcement under the United Nations Convention on the Law of the Sea (UNCLOS) and other International Maritime Organization (IMO) agreements.

The interrelationships among these roles are illustrated in Figure 2.1.

Naval vessels cannot operate across the full gamut of diplomatic and constabulary roles. A balance must be struck. The blue ellipse in Figure 2.2 indicates the balance between U.S. Navy diplomatic and constabulary roles under TSC as perceived by the U.S. Navy.[7] It indicates, for example, that vessels engaged in TSC will emphasize coalition building and avoid outright coercion.[8] It further indicates a constabu-

---

[5] Navies of interest are characterized in more detail later in this report.

[6] We have updated the roles from those described by Eric Grove to reflect international development since 1990. Principally, this means the removal of Cold War East–West confrontation as the dominant factor in the military role. See also Geoffrey Till, *Sea Power: A Guide for the Twenty-First Century*, London: Frank Cass Publishers, 2004.

[7] For clarity, the military role of the small ship is not treated in this figure.

[8] The *Joint Doctrine Encyclopedia* defines a coalition as an ad hoc arrangement between two or more nations for common action. Coalitions are typically formed on short notice and can include forces not accustomed to working together. Later, it notes that coalitions require sig-

Figure 2.1
Use of the Sea Triangle

RAND *MG698-2.1*

lary role entailing a low level of peacekeeping activities and avoiding defense of good order.[9] The constabulary role is explored in the next section.

The military role of the U.S. Navy's small combatant under TSC must build upon limited capabilities such as self defense. It is conscribed by joint doctrine as described previously in Joint Publication 3-57.1.

The U.S. Navy's view of TSC is not necessarily that of potential partner nations. To illustrate, we saw Navy missions related to amphibi-

---

nificant coordination and liaison. Differences in language, equipment, capabilities, doctrine, and procedures are some of the interoperability challenges that mandate close cooperation (*Joint Doctrine Encyclopedia*, Washington, D.C.: Joint Staff, July 16, 1997, pp. 123–125).

[9]   The DoD defines *peacekeeping* as military operations undertaken with the consent of all major parties to a dispute, designed to monitor and facilitate implementation of an agreement (ceasefire, truce, or other such agreement) and support diplomatic efforts to reach a long-term political settlement (Joint Publication 1-02, 2001, as amended through 2007). *Defense of good order* entails enforcement of international laws of the sea, largely codified in the UNCLOS. With the United States not a signatory to the UNCLOS, the U.S. Navy does not expect to enforce it.

Figure 2.2
Traditional Navy Balance Between Diplomatic and Constabulary Roles in TSC

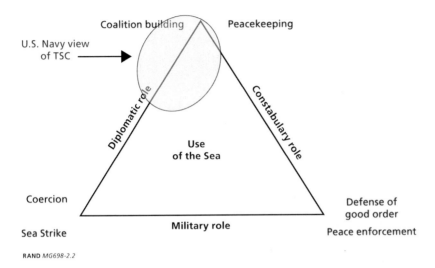

RAND *MG698-2.2*

ous warfare at the top of Table 2.1.[10] Smaller foreign navies often struggle to be effective in the constabulary role, starting within their territorial waters and then extending out to the boundaries of their exclusive economic zone (EEZ).[11] As shown in Table 2.2, many of these fleets are purely constabulary and are not intended to fight. No country of interest identified by the U.S. Navy for this study can contemplate amphibious warfare. Moreover, not all nations perceive the War on Terror in the same way as the United States, and are therefore unlikely to have the same priorities as the U.S. Navy. For the U.S. Navy, identification and interdiction of terrorists might have the highest priority. For potential partner nations, whatever their effectiveness, continuous

---

[10] Briefly, amphibious warfare is the utilization of naval firepower, logistics, and strategy to project military power ashore.

[11] The United States Coast Guard (USCG) undertakes those activities generally associated with the constabulary role. This does not mean that either force cannot undertake activities in the other roles, however. In fact, there are examples where this currently occurs: The U.S. Navy supports counter-drug operations in the Caribbean and the USCG conducts anti-piracy operations off the Horn of Africa.

control of fisheries or combating piracy may be more important. Their
view of TSC is illustrated in Figure 2.3.

Here, the constabulary role for the U.S. Navy would expand con-
siderably beyond that currently envisaged by the U.S. Navy, with diplomatic and military roles unchanged.

The difference between these two views of TSC suggests that the
U.S. Navy should also consider the needs, perhaps even the hierarchy
of needs, of potential partner nations as it plans for a small ship to conduct TSC operations. The small ship must have a constabulary capability together with its TSC capabilities.

We now start to reconcile U.S. Navy TSC objectives with the
needs of potential partner nations. For the U.S. Navy, the small ship
should be capable of supporting the War on Terror through its diplomatic role and elements of the constabulary role. Partner nations will
wish to work with the U.S. Navy to meet their hierarchy of needs,
which lie almost exclusively within the constabulary role.[12]

Figure 2.3
Potential Partner Nation Balance Between Diplomatic and
Constabulary Roles in TSC

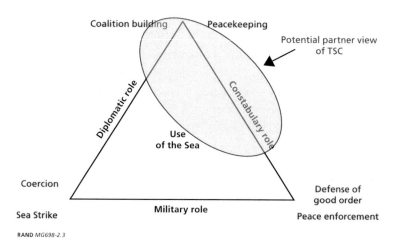

RAND MG698-2.3

---

[12] Our ranking of navies (Table 2.2) hints at the hierarchy of needs of potential partner
nations. Figures 2.2, 2.3, and 2.4 provide an explicit hierarchy of needs. Appendix B provides a more detailed hierarchy of needs.

Constabulary Role

The constabulary role centers around the requirements of policing territorial waters. The ability to operate in shallow water or from minor ports, for example, is central to this role. Rights and responsibilities of nations have grown significantly beyond the immediate distances off shore that originally formed territorial waters, and the activities of a maritime force go beyond simple police actions. Maritime nations now enjoy access to an EEZ out to 200 nautical miles and have rights and responsibilities enshrined in the UNCLOS. Within their enlarged EEZs, maritime nations may exploit natural resources, use the sea for transportation, and explore it and the seabed to understand more about its history, heritage, and potential. Maritime nations have responsibilities to maintain sovereignty, ensure freedom of navigation for others, protect the environment, and help those in distress. These rights and responsibilities have been characterized as *good order at sea*.[13]

In order to secure their rights, maritime nations must maintain a presence at sea and be prepared to enforce good order with police operations. Presence is a passive activity; it might be fleeting, though with acceptable revisit times, and undertaken from the air by maritime patrol, or it might be permanent with, for example, surveillance of busy straits from land stations or ships at sea on patrol. In this sense, maritime forces are reactive; their presence allows them to notice events requiring action. Tasks and their associated capabilities for maritime presence are shown in Figure 2.4. These tasks and capabilities are hierarchical; maritime forces must be able to conduct maritime patrols before they can conduct search and rescue operations, IMO obligations operations, and environmental protection. In turn, they must be able to conduct search and rescue operations before they can conduct IMO obligations operations and environmental protection operations. Finally, they must be able to conduct IMO obligations operations before they can conduct environmental protection operations.

---

[13] Till, Chapter Ten.

Figure 2.4
Tasks and Capabilities for Maritime Presence

Tasks				Capability
Maritime Patrol	Search and Rescue	IMO Obligations	Environmental Protection	
				Operate freely in shallow water
				Operate from minor ports
				Operate in difficult sea states
				Operate for long periods—endurance; replenishment at sea and remote ports
				Operate for long periods—system redundancy; at sea and remote port maintenance
				Operate for long periods—crew fatigue
				Conduct accurate and redundant navigation
				Detect and locate those in distress
				Perform small-scale rescue missions
				Contribute to large and dispersed rescue missions
				Provide first aid (personnel and materiel assistance)
				Update hydrographic information
				Update weather information
				Contain small-scale pollution incidents
				Contribute to containment of large-scale pollution incidents
				Detect different types of pollution and identify sources

RAND *MG698-2.4*

Police operations (shown in Figure 2.5) as a class are proactive, but they build on passive presence activities; maritime forces capable of police operations must first be able to perform the tasks and have the capabilities for maritime presence. As with the tasks and capabilities for maritime presence operations, these tasks and capabilities are again hierarchical. Here, maritime forces must be capable of enforcing the right of innocent passage[14] before they can protect the maritime infrastructure, and so on as above.

_____

[14] The right of innocent passage in the territorial sea entitles foreign ships to navigate through those waters. Such passage may include stopping and anchoring insofar as they are incidental to ordinary navigation or are rendered necessary by *force majeure* or distress. Passage is deemed innocent so long as it is not prejudicial to the peace, good order, or security of the coastal state. The UNCLOS list of non-innocent activities includes fishing and acts of

Figure 2.5
Tasks and Capabilities for Police Operations

Tasks			Capability
Enforcement of Innocent Passage	Protection of Maritime Infrastructure	Economic and Law Enforcement	Detect and track local surface contacts
			Detect and track wide area surface contacts
			Identify surface contacts
			Contribute to recognized operational picture
			Intercept surface contacts
			Build operational intelligence picture
			Operate in anchorages, port approaches, and ports
			Counter attack-swimmers
			Counter improvised small boat attacks
			Assist in explosive ordnance disposal operations
			Engage hostile surface contacts
			Conduct boarding operations
			Escort detained vessels
			Detain and transfer high-value prisoners
			Conduct fishery and smuggling enforcement operations
			Interdict pirate and terrorist vessels

RAND MG698-2.5

In addition to the constabulary role missions described above, the U.S. Navy vessel will also need to be able to undertake another set of tasks and capabilities. Those include the TSC tasks that lie within the diplomatic role. Again presented hierarchically, they are foreign visits, in-country training (on land or at sea), combined exercises, and combined operations.[15] These tasks and their associated capabilities are shown in Figure 2.6. In this hierarchy, for example, the U.S. Navy

---

willful and serious pollution. Coastal states may take whatever steps are necessary to prevent passage that is not innocent, and may thus exclude such vessels from the territorial sea.

[15] *Combined* indicates multinational involvement—i.e., U.S. Navy and partner nation vessels. In the capabilities list, *regional* refers to the involvement of other local nations in addition to the primary partner nation. Thus, the most demanding activity is a regional combined operation that involves multiple nations—i.e., the U.S. Navy with one or more partner nations and other local nations.

vessel must be prepared to operate in shallow water and from minor ports before it can conduct in-country training.

An extended tasks and capabilities list that brings together Figures 2.4–2.6 is shown in Appendix B. It is analogous to the N816 task list in that it describes the effects of the N816 tasks within the construct of the universal Use of the Sea concept.

Figure 2.6
Tasks and Capabilities for the TSC Mission

Tasks				Capability
Foreign Visits	In-Country Training (Land/Sea)	Combined Exercises	Combined Operations	
				Operate freely in shallow water
				Operate from minor ports
				Open ship to visitors
				Conduct VIP ship tours
				Engage in local community activities (such as sports)
				Engage in civil activities (such as fundraising and local small-scale engineering or infrastructure projects)
				Engage in professional discussions and future combined activity planning
				Conduct shore-based training—small ship's staff
				Conduct shore-based training—additional teams supporting small ship's staff (or vice versa)
				Conduct alongside training—U.S. Navy and foreign nation's ships
				Conduct underway training—U.S. Navy and foreign nation's ships
				Conduct combined operations—single ship
				Conduct combined operations—multiple ship
				Conduct combined operations—regional
				Conduct combined exercises—single ship
				Conduct combined exercises—multiple ship
				Conduct combined exercises—regional

The Small Ship Concept of Employment

The small ship concept of employment presented here was developed iteratively and concurrently with our examination of the presence and police operations missions of the constabulary role together with the requirements of the TSC mission. Simultaneously, we analyzed the N816 task list and its implied capabilities. As noted previously, the original N816 list included tasks that a small vessel will never undertake, such as providing area air defense. The provided list was resolved alongside our own to allow comparison and as a crosscheck for completeness.

The concept of employment evolved into a limited role that requires a single small vessel or squadron of vessels that rely on support from ashore or afloat assets and operate in a low military threat environment. As discussed above, although the vessel will be supporting the War on Terror, it will do so through TSC and execution of the constabulary role as allowed by the partner nation. Its concept of employment can be further defined through the following primary and linked missions:

- **Primary mission.** Full-scope TSC engagement with developing nations at the periphery of traditional U.S. Navy influence. Tasks include military liaison, training of maritime forces of partner nations, and acting as a positive influence in support of good order missions of partner nations.
- **Linked missions.** Intelligence, surveillance, reconnaissance, maritime interception operations, maritime support operations, maritime domain awareness, antiterrorism/force protection, homeland security, special operations, law enforcement, civil affairs, and humanitarian assistance.

CHAPTER 3

Linking Tasks to Capabilities

The previous chapter defined tasks and first-order capabilities required of the small ship under the N816 task list and for the task list developed by RAND for this study. We now refine the following first-order capabilities shown in Figures 2.1–2.6 across both task lists:

- **Operate freely in shallow water**. The small ship will need to operate close to shore in parts of the world where oceanographic information may be limited or based on old surveys. A vessel constrained to deep water because of fears of grounding would be limited in its effectiveness, particularly in the close to shore coastal areas where presence and police operation missions might be needed.
- **Operate from minor ports**. Similarly, many ports in Africa and other parts of the world are poorly surveyed. Furthermore, channels are not maintained by dredging or placement of navigational marks. In such circumstances, the capability to operate in shallow water is important. Additionally, less well-developed ports can demand more of a vessel: Jetty or quay facilities might be non-existent or "rough and ready," and ports may lack normal harbor support services such as tugs, cranes, and brows. A lack of facilities places additional requirements on vessels wishing to operate safely from these minor ports.
- **Operate in difficult sea states**. Sea conditions will directly affect the safety and mission effectiveness of a small vessel. Wind strength and direction, wave height and periodicity, swell, and

other factors can all combine to make small vessel operation difficult. Upon a review of National Oceanic and Atmospheric Administration wave height data, we found that sea state considerations do not put a meaningful lower bound on vessel size. However, operations in some of the indicated sea states would be difficult for any small vessel. Appendix A provides material on this matter.

- **Operate for long periods—endurance.** This capability addresses the physical constraints imposed by the small size of the vessel. Limited space for storing fuel, water, food, and other consumable supplies will restrict how long a vessel can remain at sea. Where circumstances and the design of a vessel allow, replenishment of these supplies can occur at sea. Alternatively, it may be possible to replenish from remote ports other than those where the vessel is temporarily based for the current part of the deployment.

- **Operate for long periods—system redundancy.** Ships and their systems need continuous monitoring and occasional inspection. Systems do fail. In a small vessel, where space is already at a premium, it is more difficult to carry the specialized engineering personnel and range of stores needed for technical maintenance. Vessel and system design can mitigate these concerns, as can the use of redundant essential systems. Since the small vessel will be operating in remote regions of the world where outside assistance may be many days away, the ability to return safely to port is important.

- **Operate for long periods—crew fatigue.** Because a small vessel will require a small crew, there is unlikely to be sufficient space for the extra watch-standers that larger vessels might carry. Consequently, the crew will likely work longer hours. The motion of a small vessel will also hasten fatigue. The rudimentary crew facilities of most small vessels will make it more difficult for personnel to recuperate when off watch.

- **Operate in anchorages, port approaches, and ports.** Enforcement of good order will need to be undertaken not only at sea, but also where merchant vessels and other maritime traffic congregate. Typically, these are places such as anchorages, port approaches, and ports. These are places where presence is important to identify possible illegal activity and to protect innocents from attack. Small vessels tend to be well-suited for operations in these confined waters.

- **Develop surface picture.** This key capability draws together a number of individual capabilities that together lead to actionable picture compilation. These capabilities include the detection, tracking, and identification of surface contacts, and potentially the processing of intelligence information, which together would allow a vessel to proactively investigate contacts of interest.
- **Conduct boarding/detachment operations.** In most situations, the ability to board a vessel of interest will be the only way to identify its purpose and enforce compliance with relevant laws. For vessels already determined as hostile, armed boarding operations may be required to detain the vessel and its crew. For this small vessel, the boarding capability represents its main "weapon system."
- **Engage hostile contacts.** The small vessel will need the capability to engage suspect or hostile craft in a variety of scenarios (e.g., to preclude the need to board, to make craft accept a boarding, or to support an armed boarding). The system will also need to be able to fire warning shots and then escalate to accurate disabling fire so as to remain compliant with rules of engagement and other enforcement restrictions.

CHAPTER 4

Capabilities-to-Characteristics and Options

This chapter continues our strategies-to-tasks methodology by relating the capabilities described in the previous chapter to vessel characteristics. Constraints, tradeoffs between desired capabilities, and obvious balance points are identified. Many of the key capabilities relate to the primary characteristics of the small ship in a complex many-to-many manner as depicted in Figure 4.1.

Figure 4.1
Many-to-Many Relationship Between Capabilities and Characteristics

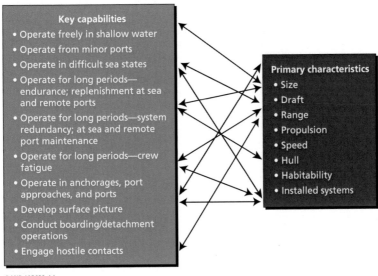

Key capabilities
- Operate freely in shallow water
- Operate from minor ports
- Operate in difficult sea states
- Operate for long periods—endurance; replenishment at sea and remote ports
- Operate for long periods—system redundancy; at sea and remote port maintenance
- Operate for long periods—crew fatigue
- Operate in anchorages, port approaches, and ports
- Develop surface picture
- Conduct boarding/detachment operations
- Engage hostile contacts

Primary characteristics
- Size
- Draft
- Range
- Propulsion
- Speed
- Hull
- Habitability
- Installed systems

RAND *MG698-4.1*

Such relationships cannot be finally determined without considering detailed designs, which would be beyond the scope of this study. Instead, study analysts adopted a qualitative approach, relying upon insights from interviews with shipbuilding industry and naval experts experienced in small vessel design and operation, together with the assessments of subject matter experts on the study team.[1]

Most of the key capabilities were found to relate to hull and mechanical factors. Vessel draft, for example, relates directly to the capability to operate in shallow water and from minor ports. However, shallow draft can lead to poor sea-keeping qualities, which, in turn, can increase crew fatigue and reduce endurance. The small vessel characteristics identified were size, draft, range, propulsion, speed, hull, habitability, and installed systems. We discovered some of the following trades in the many-to-many relationship:

- The capability to operate freely in shallow water (water with a depth of 15 feet or less) is driven primarily by vessel draft.
- The capability to operate from minor ports (i.e., ports with navigable depth of 15 feet or less) also depends on draft as well as propulsion type. There is a preference here for steel hulls rather than aluminum or fiberglass for ruggedness.
- The capability to operate in anchorages, port approaches, and ports is similar to that for capability to operate from minor ports. It places a greater emphasis, however, on vessel size for maneuverability.
- The capability to operate in difficult sea states (roughly sea states 3 and above, depending on the complexity of wave action) relates to size, draft, and hull design.
- The capability to operate for long periods relates to factors of endurance, including available quantities of consumables, reliability/maintainability, and crew fatigue. Endurance, in turn, depends on range, hull design, and propulsion power. Reliability

---

[1]   CDR David Newton, U.S. Coast Guard, and LCDR Michele Poole, U.S. Navy, are surface warfare officers.

and maintainability depend on propulsion type (with diesels preferred) and redundancy. Crew fatigue relates to hull design, hull size, and habitability.

- The capability to intercept surface contacts depends primarily on speed, which goes back to hull size and propulsion. In this instance, bigger is not always better.
- The capability to conduct boarding operations is driven primarily by size (here, bigger is better) and by hull vulnerability (e.g., risk of exposure to enemy action or accidental collisions).
- The capability to detect, track, and engage hostile contacts, for which installed systems are most important.

These competing demands relate primarily to presence (and hence to the constabulary role). With a certain level of constabulary competence, depending upon the size of the vessel, it is possible to incorporate some or all of the TSC capabilities that demand space and a larger crew. More constabulary capability could also be added to larger vessels. On the other hand, if the space available on a small ship is inadequate for TSC-related capabilities, a support ship could be used. A continuum of small ships, in combination with support ships, is possible.

Emergence of Three Vessel Classes

Balancing the key capabilities and related characteristics, and then considering the remaining capabilities required, three classes of small vessel, defined both by their size and the level of support they would need,[2] emerged—nearshore, coastal, and offshore:

- **Nearshore patrol vessel.** These vessels nominally displace fewer than 100 tons. They require logistic and operational support, including the following: hotel services; refueling, rearmament,

---

[2] Support could be provided from sea or shore. To simplify our discussion, we now concentrate on sea support in the form of a mothership. A ship would provide greater flexibility and reduce demands on the partner nation. We have not looked in detail at any potential shore support options, however.

and re-supply; additional rotational crews; maintenance facilities and support; feed of situational awareness;[3] and provision of additional command, control, communications, computers, and intelligence (C4I) support.[4] *These vessels would require a specialized mothership.*

- **Coastal patrol vessel.** These vessels nominally displace between 300 and 700 tons. They require some logistic and operational support, including regular refueling and re-supply, some situational awareness, and tailored C4I support. *These vessels would benefit from a mothership of opportunity.*
- **Offshore patrol vessel.** These vessels nominally displace about 1,500 tons. They would benefit from logistic and operational support, including occasional refueling and re-supply, some situational awareness, and tailored C4I support. *No mothership would operate with these vessels.*

---

[3]   A nearshore patrol vessel, particularly one envisaged as a cheap, simple solution suitable for use by partner nations, would not have the systems or personnel needed to compile a sophisticated operational picture. The support vessel (or shore facility) would need to do this.

[4]   Again, a very small vessel will not have the personnel to undertake wide area command and control or the space for enhanced communications and intelligence equipment.

CHAPTER 5

Vessel Survey and Assessment

World Survey

We considered more than 1,000 vessels from over 60 nations in our world survey.[1] Unsurprisingly, the survey yielded three groupings of vessels that aligned closely in size (tonnage) with the three classes of vessel we identified previously. Figure 5.1 lists six vessels that are representative of the database survey findings and that broadly meet the capabilities of the three classes identified during our theoretical approach.

Additionally, the study team approached several European navies to discern how they meet the requirements for the types of operations envisaged for the small vessel. France and the United Kingdom, which both undertake constabulary roles within the EEZs of their overseas territories, were most helpful. France uses two classes of vessel in TSC: the Floreal frigate and the P400 Class. Floreal frigates are built to merchant passenger marine standards. They displace 3,000 tons, have a flank speed of 20 knots, host 80 crewmembers, and can carry an embarked helicopter. Their capabilities include traditional roles such as anti–surface warfare and they have a limited capacity for anti–air warfare. France has sold two of these frigates to Morocco. P400 Class patrol boats were designed for police and customs operations from overseas territories. They displace 450 tons, have a flank speed of 24

---

[1] This survey drew primarily upon the Haze Gray Web site (http://www.hazegray.org) and Jane's Intelligence Service (available by subscription only). Survey results were crosschecked against material provided by OPNAV N81.

Figure 5.1
Examples of Three Vessel Classes

	Country	Ship class	Tons	Draft (ft)	Crew size	Diesels/ shafts	Speed (kt)	Range (nm)
Offshore	United Kingdom	River Class	1,550	11.2	30–42	2	20	10,000
	United States (USCG)	Famous Class	1,780	14.1	100	2	19	12,700
Coastal	United States (USCG)	Fast Response Cutters	325	8.2	18–20	2	30	4,230
	Ireland	Orla (Peacock) Class	712	8.9	39	2	25	2,500
Nearshore	Cameroon	Rodman 101	63	5.9	9	2	26	800
	St. Vincent and the Grenadines	VT 75 ft	70	7.9	11	2	24.5	1,000

RAND *MG698-5.1*

knots, and host 24 crewmembers. P400 Class patrol boats have been sold to Gabon and Oman.

The United Kingdom has recently employed an innovative approach to procuring the new generation of River Class offshore patrol vessels (OPVs)[2] to enhance capability and reduce cost. Under this approach, vessels are owned by the shipbuilder and leased for an initial period of five years along with worldwide contractor logistic support. The leasing contract requires 960 operational days per year from three vessels with 1,040 days of operation achieved. River Class OPVs are built to commercial shipbuilding standards. Contractor logistic support considerations are incorporated into the ship design to improve endurance and redundancy and to ease maintenance. These considerations added 4 percent to the shipbuilder's costs.

Both France and the United Kingdom consider complementary airborne maritime surveillance very important.

---

[2] The original River Class OPV is a flexible design. The latest Royal Navy version of the River Class vessel has an enhanced design with a flight deck and limited aviation facilities.

Recent Acquisitions by Navies of Interest

To take our analysis of current vessels further, and to better understand the needs of potential partner nations, the vessels ordered recently by 21 countries of interest were surveyed in detail using unclassified sources. The results are displayed in Figure 5.2, which shows existing naval vessels and (in the box on the right) future naval vessels that are currently under contract. As can be seen, these countries seem to need a group of vessels around the 300–500 ton mark. A growing need for much larger vessels, around 1,500 tons and up, is also evident. As noted earlier, this is likely to be a consequence of the accession of more and more countries to the UNCLOS, which has increased the requirement for effective maritime forces that are able to undertake the constabulary role throughout much larger EEZs.

Figure 5.2
Survey of Recently Delivered or Anticipated Naval Vessels for 21 Countries of Interest

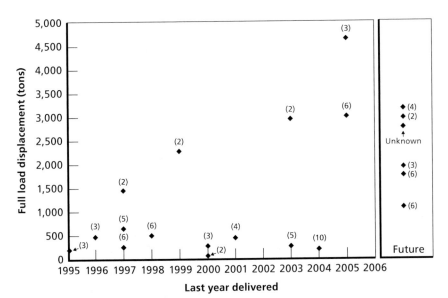

NOTES: The identities of these 21 countries are considered sensitive. Numbers in parentheses indicate the number of vessels delivered.

RAND MG698-5.2

We concluded from these surveys that the three classes of vessel derived from our strategies-to-tasks analysis were realistic because many examples of such vessels have been built, are at sea, and are practical, since such vessels are being used in the constabulary role. We also confirmed that our three ship classes bound the options of such vessels.

Assessment of Representative Vessels

With survey information and a clearer understanding of the actual capabilities of the vessels in each class, we qualitatively judged how well they might perform against our constabulary and TSC capability lists. We also wanted to reconcile capabilities implied by the N816 task list with our own. A simple stoplight system was used to assess the performance of each class against the identified capabilities in each list. The assessment excluded any contribution to capabilities that a mothership might make for the following reasons: We had used the level of mothership support together with small ship size to define the vessel classes and needed to avoid circular reasoning; any hypothetical mothership capability could fill in any small vessel gaps and make the assessment pointless (there would be full compliance for each class); and this was a study about small vessels and we had not researched or analyzed potential mothership capability.[3]

The qualitative assessments were made using red, yellow, or green to indicate the extent of any discrepancy or to indicate the degree of capability compliance:

- red—critical disqualifying discrepancy; least compliance
- yellow—moderate discrepancy; partial compliance
- green—no discrepancy; full compliance.

---

[3] We reviewed several current and potential examples of a mothership to ensure that those capabilities not inherent in the small vessel could be placed in a support platform. The Royal Navy's *RFA Diligence* is one such ship, and other suggestions included a modified Joint High Speed Vessel or Littoral Combat Ship. The U.S. Navy and USCG have used support ships of varying capabilities in mothership roles.

This system was used to evaluate small ships independent of a mothership. However, the system applies equally to the mothership in that it describes what the mothership would have to provide to achieve needed capabilities.

In Figure 5.3, we show assessments against the RAND capability list. In Figure 5.4, we show simplified versions of the assessments side by side. As can be seen in both figures, there is a distinct trend for each class of vessel that is mirrored in both capability lists. This reinforces our confidence in our methodology and the objectivity of our judgments.

As stated previously, these assessments, conducted without consideration of a mothership, reflect upon both the small vessel and the mothership. For the nearshore patrol vessel, each figure shows roughly 20 critical disqualifying discrepancies or instances of least compliance. For the coastal patrol vessel, each figure shows few or no disqualifying discrepancies or instances of least compliance. For the offshore patrol vessel, each figure shows only a few discrepancies, and demonstrates partial or full compliance in all instances. Upon further scrutiny, we found that the causes of discrepancies (e.g., inadequate endurance) for each class were the same for both lists. Accordingly, we concluded that the two task lists, in aggregate, are interchangeable. This resolves the problem of two distinct task lists in our strategies-to-tasks methodology.

Cost Estimates

We took our comparison one stage further by considering potential rough order of magnitude costs for each solution:[4]

- Nearshore patrol vessel (<100 tons)
  - Example: USCG Marine *Protector* Class coastal patrol boat

---

[4] Program vessel costs came from USCG for *Protector* Class and *Deepwater* and VT Group for the River Class OPV.

- Cost: approximately $5 million (fully outfitted), plus approximately $200 million for a specialized mothership (notionally, a Joint High Speed Vessel)
- Displacement: 90 tons
- Length: 87 feet
- Draft: approximately 6 feet
- Crew: approximately 10

• Coastal patrol vessel (300–700 tons)
- Example: USCG *Deepwater* Fast Response Cutter
- Cost: approximately $40 million (fully outfitted)
- Displacement: 325 tons
- Length: 140 feet
- Draft: approximately 8 feet
- Crew: 16–18

• Offshore patrol vessel (>1,500 tons)
- Example: UK River Class OPV
- Cost: approximately $50 million (built to commercial specifications with options meeting requirements)
- Displacement: approximately 1,550 tons
- Length: 262 feet
- Draft: approximately 11 feet
- Crew: 30–42.

Figure 5.3
Comparative Assessment of Vessel Classes Using RAND's Capability List

Mission		Task	Capabilities	Nearshore patrol vessel	Coastal patrol vessel	Offshore patrol vessel
Maritime presence		Patrol	Operate freely in shallow water			
			Operate from minor ports			
			Operate in difficult sea states			
			Operate for long periods—endurance; replenishment at sea and remote ports			
			Operate for long periods—system redundancy; at sea and remote port maintenance			
			Operate for long periods—crew fatigue			
			Conduct accurate and redundant navigation			
		Search and rescue	Detect and locate those in distress			
			Perform small-scale rescue missions			
			Contribute to large and dispersed rescue missions			
			Provide first aid (personnel and materiel assistance)			
		IMO obligations	Update hydrographic information			
			Update weather information			
		Environmental protection	Contain small-scale pollution incidents			
			Contribute to containment of large-scale pollution incidents			
			Detect different types of pollution and identify sources			
	Police operations	Enforcement of innocent passage	Detect and track local surface contacts			
			Detect and track wide area surface contacts			
			Identify surface contacts			
			Contribute to recognized operational picture			
			Intercept surface contacts			
			Build operational intelligence picture			
		Protection of maritime infrastructure	Operate in anchorages, port approaches, and ports			
			Counter attack-swimmers			
			Counter improvised small boat attacks			
			Assist in explosive ordnance disposal operations			
			Engage hostile surface contacts			
			Conduct boarding operations			
		Economic and law enforcement	Escort detained vessels			
			Detain and transfer high-value prisoners			
			Conduct fishery and smuggling enforcement operations			
			Interdict pirate and terrorist vessels			
	TSC	Foreign visits	Operate freely in shallow water			
			Operate from minor ports			
			Open ship to visitors			
			Conduct VIP ship tours			
			Engage in local community activities (such as sports)			
			Engage in civil activities (such as fundraising and local small-scale engineering or infrastructure projects)			
			Engage in professional discussions and future combined activity planning			
		Combined training (land/sea)	Conduct shore-based training—small ship's staff			
			Conduct shore-based training—additional teams supporting small ship's staff (or vice versa)			
			Conduct alongside training—U.S. Navy and foreign nation's ships			
			Conduct underway training—U.S. Navy and foreign nation's ships			
		Combined exercises	Conduct combined exercises—single ship			
			Conduct combined exercises—multiple ship			
			Conduct combined exercises—regional			
		Combined operations	Conduct combined operations—single ship			
			Conduct combined operations—multiple ship			
			Conduct combined operations—regional			

Figure 5.4
Comparative Assessment of Vessel Classes Using the RAND and N816 Capability Lists

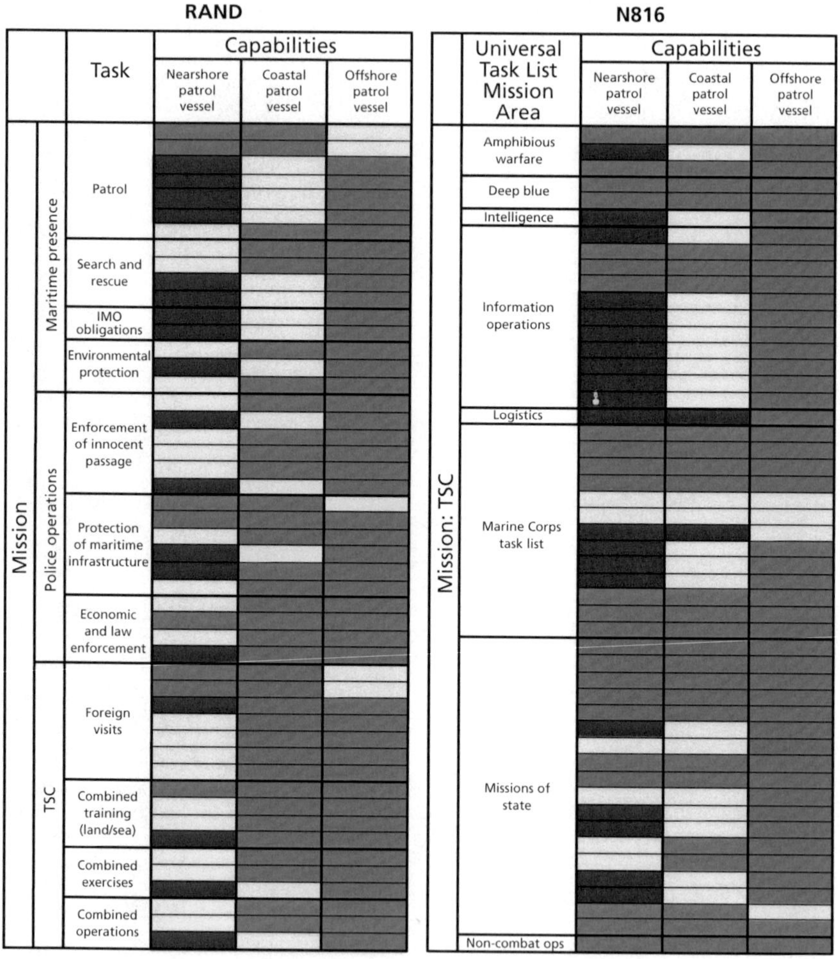

RAND MG698-5.4

CHAPTER 6

Findings, Observations, and Next Steps

Overall Findings

Each vessel class comes with strengths and weaknesses. The following paragraphs summarize key points for each class.

The nearshore patrol vessel is the smallest and least-expensive vessel with greatest access to shallow waters and minor ports. The low unit procurement cost would be offset for the U.S. Navy, however, by the need for a dedicated, specialized mothership to support this vessel. The nearshore patrol vessel has the worst habitability, would be the least survivable in rough seas or because of enemy action, and would be the least-capable vessel. Finally, while it is an attractive entry-level vessel to some nations, potential partner nations are now buying larger vessels.

The coastal patrol vessel is also a small vessel with good access to shallow waters and minor ports. Its increased size gives this vessel numerous operational advantages over the nearshore vessel, including better survivability, greater endurance, and improved habitability. Coastal patrol vessels would not need a specialized mothership and could be supported by a suitable vessel of opportunity, although such a ship might not provide as much support as a specialized mothership. The larger size of the coastal patrol vessel would enable it to work more comfortably with the relatively larger vessels (over 1,000 tons) now being purchased by potential partner nations.

The offshore patrol vessel is the largest of the vessels considered in this study. It would have the greatest independence of operation and is the most capable and versatile vessel, able to undertake long patrols. In size, it is the most comparable to the larger vessels being purchased by potential partner nations. With these advantages comes increased cost, however. A fixed budget would allow fewer to be purchased, potentially reducing regional presence.

Study Team Observations

In this quick survey report on small ships, the RAND team did not draw definitive conclusions; rather, we present several observations for the U.S. Navy to consider in a more definitive study of the small ships phenomenon and their employment in the War on Terror. First, as suggested in the earlier chapters, the U.S. Navy needs to give more consideration to the constabulary needs of potential partner nations in order to gain the increased access it needs to undertake TSC (and, potentially, operations in support of the War on Terror).

Second, whichever small vessel is chosen, success in TSC will greatly depend on the qualities of the crew. Small ship personnel will need to be specialized (with language training, for example) and they will require stability in their assignments to assure adequate time in theater and to prevent untimely personnel rotations. In this way, they will stand a better chance of building long-term relationships with the navies of partner nations. We recommend that the crews of these vessels be specially selected, with skills akin to those of SOF teams, and that they be given specific training to improve their abilities in constabulary and TSC tasks.

Third, while specialized motherships offer advantages in terms of operational suitability, they also have the disadvantage of needing to be procured through a formal acquisition process that could delay the implementation of the small vessel concept. In addition, such vessels might have limited utility in wider combat operations and would represent significant additional cost to the U.S. Navy.

Fourth, there is value in quantity. Having many of these small ships would be more beneficial to the U.S. Navy's concept of a vessel for the War on Terror than fewer. A squadron of five ships, for example, could support each other and provide the necessary intelligence and command, control, communications, computers, intelligence, surveillance, and reconnaissance (C4ISR) meshing that a larger ship alone could not.

Fifth, most countries of interest are gravitating toward larger rather than smaller ships. If the U.S. Navy wants to operate with those navies in the future, a comparable ship should be considered. If the U.S. Navy wishes to sell such a ship through the U.S. Department of State's Foreign Military Sales program, the larger ship again seems to be of more interest to potential partner nations (based on their recent ship procurements).

Sixth, while motherships (i.e., Navy combatants especially configured to support one or more small ships) offer advantages, they are costly and potentially vulnerable because they probably could not access the same ports as the ships they are designed to support.

Finally, the study team does not recommend the nearshore patrol vessel. Considerations of fuel, stores, and crew fatigue give it the least endurance of the three potential solutions. Crew fatigue will be exacerbated by the small crew size (approximately ten persons), which will also require steady crew rotation and the need for multiple crews in theater. It has the worst habitability, especially in difficult sea states, and is least survivable in terms of both seakeeping and vulnerability to small arms fire. It was seen as least capable of performing the overall constabulary and TSC missions, and would be most dependent on a dedicated mothership for capability.

This study is a preliminary analysis of new or changing missions that the U.S. Navy may face as it attempts to partner with maritime nations beyond those with which it has enjoyed longstanding relationships. It was not possible in this small study to be as definitive in our research and analysis as we might desire. Additionally, the U.S. Navy may wish to test and develop our concept of employment, looking at issues such as how to get the small vessels into a theater and support them once deployed. Other key steps might include determining the force structure of the mothership (number and type) and the small vessels. Finally, small vessels may be able to contribute to missions beyond those of constabulary and TSC.

Possible Next Steps for the U.S. Navy

We suggest that the U.S. Navy validate and further develop a concept of employment for the small ships. The concept of employment should consider

- how the small ship reaches the theater of interest
- how the small ship is supported in theater, including
  - the potential use of contractor logistic support
  - new manning options, including longer tours for crew
  - the concept of the mothership and its use, including cost, basing rights, load lists, etc.
  - the possibility of partner nation support, including the potential advantage of working in-country with a host nation and the potential disadvantage to force protection
  - issues of force structure, including the merits of squadrons for combined operations and the issue of support if the ships are not homeported or based in a host nation.

Finally, the potential roles for a small ship outside the TSC world should be examined.

CHAPTER 7

Epilogue

Shortly after this analysis was briefed out to the U.S. Navy, the service was directed to examine the PC-1 Cyclone Class (shown in Figure 7.1) as a small ship for use in TSC. For use in TSC, the PC-1 was to be given an updated propulsion system and improved command and control for greater connectivity. A non-stabilized 25-mm gun was to be replaced by a stabilized 25-mm gun. The PC-1 displaces 331 tons, placing it toward the low end of the coastal patrol vessel band (300–700 tons).

Figure 7.1
USS Cyclone, PC-1

SOURCE: U.S. Navy photo.
RAND *MG698-7.1*

At the U.S. Navy's request, the RAND study team conducted a short follow-on study of the PC-1 using data on these updates and improvements and applying the methodology previously described. We found that the PC-1 would be somewhat less capable than the notional coastal patrol vessel used in this study, but that mothership support would render it fully capable. This result was accepted for use by the U.S. Navy.

Environmental Analysis

Study analysts initially suspected that the smallest ships considered in this study, nearshore patrol vessels, might be unable to operate in some environments of interest. This hypothesis was tested using a wave height database from the Asheville, North Carolina, detachment of the Fleet Numerical Meteorology and Oceanography Center (FNMOC).[1] The FNMOC database indicates the frequency of occurrence of waves of given heights (measured from trough to crest) for given locations worldwide.

We were not able to translate material in the FNMOC database into sea states, which are described in terms of significant wave height.[2] Instead, we identified cutoff wave heights for the highest 10 percent of waves and as indicative of sea conditions. For example, if the cutoff for the tenth percentile of waves is 6 feet, then 90 percent of the waves are below 6 feet and 10 percent of the waves exceed 6 feet in height. Results are presented using contour plots using Mathematica, a commercial software tool.

This procedure enabled us to identify the most severe season for each region. For example, Figure A.1 below shows tenth percentile wave heights around Africa in July, the most severe weather season in this region. Directed to focus on sub-Saharan Africa, we concluded that nearshore patrol vessels could operate within EEZs of interest off

---

[1] The database, *Marine Climatic Atlas of the World*, comes in a CD format and is freely available.

[2] Significant wave height is commonly defined as the average of the highest one-third of the waves.

Africa. Conditions would be fatiguing to the crews of such vessels, but the vessels would not be directly jeopardized.

Selected countries in Central America, the Caribbean, and northern South America were of additional interest to U.S. Navy planners. In this region (shown in Figure A.2), where the worst season, environmentally, also occurs in July, nearshore patrol vessels could operate safely.

Finally, some selected countries of interest to U.S. Navy planners are in the Western Pacific, especially Oceania. In this region (shown in Figure A.3), where the worst conditions prevail in January, nearshore patrol vessels could operate safely.

We conclude from this analysis that sea state conditions do not exclude any vessel class, but could impact operations of the smallest vessels through such factors as crew fatigue.

Figure A.1
Most Severe Conditions—Africa

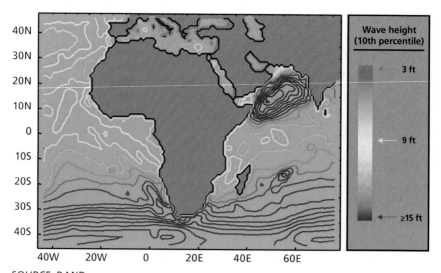

SOURCE: RAND.

RAND MG698-A.1

Figure A.2
Most Severe Conditions—South America

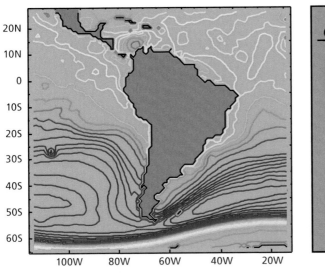

SOURCE: RAND.
RAND *MG698-A.2*

Figure A.3
Most Severe Conditions—Western Pacific

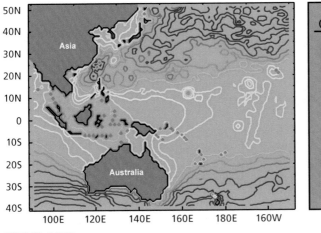

SOURCE: RAND.
RAND *MG698-A.3*

Extended Task List

The main body of this report presented selected tasks and first-order capabilities required of the small ship under an N816-provided task list and two study-generated task lists. An extended list of tasks and capabilities, combining tasks and capabilities across maritime presence, police operations, and TSC operations, is provided in Table B.1 below.

Figure B.1
Extended List of Tasks and Capabilities

Mission	Task	Capabilities
Maritime Presence / Police Operations	Patrol / Search and Rescue / IMO Obligations / Environmental Protection / Enforcement of Innocent Passage / Protect Maritime Infrastructure / Economic and Law Enforcement	Operate freely in shallow water
		Operate from minor ports
		Operate in difficult sea states
		Operate for long periods—endurance
		Operate for long periods—system redundancy
		Operate for long periods—crew fatigue
		Conduct accurate and redundant navigation
		Detect and locate those in distress
		Perform small-scale rescue missions
		Contribute to large and dispersed rescue missions
		Provide first aid (personnel and materiel assistance)
		Update hydrographic information
		Update weather information
		Contain small-scale pollution incidents
		Contribute to containment of large-scale pollution incidents
		Detect different types of pollution and identify sources
		Detect and track local surface contacts
		Detect and track wide area surface contacts
		Identify surface contacts
		Contribute to recognized operational picture
		Intercept surface contacts
		Build operational intelligence picture
		Operate in anchorages, port approaches, and ports
		Counter attack-swimmers
		Counter improvised small boat attacks
		Assist in explosive ordnance disposal operations
		Engage hostile surface contacts
		Conduct boarding operations
		Escort detained vessels
		Detain and transfer high-value prisoners
		Conduct fishery and smuggling enforcement operations
		Interdict pirate and terrorist vessels
U.S. Navy TSC	Foreign Visits / In-Country Training (Land/Sea) / Combined Exercises / Combined Operations	Operate freely in shallow water
		Operate from minor ports
		Open ship to visitors
		Conduct VIP ship tours
		Engage in local community activities (such as sports)
		Engage in civil activities (such as fundraising and small-scale infrastructure projects)
		Engage in professional discussions and future combined activity planning
		Conduct shore-based training—small ship's staff
		Conduct shore-based training—additional teams supporting small ship's staff
		Conduct alongside training—U.S. Navy and foreign nation's ships
		Conduct underway training—U.S. Navy and foreign nation's ships
		Conduct combined exercises—single ship
		Conduct combined exercises—multiple ship
		Conduct combined exercises—regional
		Conduct combined operations—single ship
		Conduct combined operations—multiple ship
		Conduct combined operations—regional

Bibliography

Department of Defense, *Building Partnership Capability: QDR Execution Roadmap*, 2006.

Department of the Navy, *Naval Operations Concept 2006*.

Department of the Navy, Fleet Numerical Meteorology and Oceanography Center, *Marine Climatic Atlas of the World*, CD-ROM, 1995.

Grove, Eric, *The Future of Sea Power*, Annapolis, Md.: Naval Institute Press, 1990.

Joint Doctrine Encyclopedia, Washington, D.C.: Joint Staff, July 16, 1997.

Joint Publication 1-02, *Department of Defense Dictionary of Military and Associated Terms*, Washington, D.C.: Joint Staff, April 12, 2001, as amended through October 17, 2007.

Joint Publication 3-57.1, *Joint Doctrine for Civil Affairs*, Washington, D.C.: Joint Staff, April 14, 2003.

Marquis, Jefferson P., Richard E. Darilek, Jasen J. Castillo, Cathryn Quantic Thurston, Anny Wong, Cynthia Huger, Andrea Mejia, Jennifer D.P. Moroney, Brian Nichiporuk, and Brett Steele, *Assessing the Value of U.S. Army International Activities*, Santa Monica, Calif.: RAND Corporation, MG-329-A, 2006. As of December 4, 2007:
http://www.rand.org/pubs/monographs/MG329/

Moroney, Jennifer D.P., Adam Grissom, and Jefferson P. Marquis, *A Capabilities-Based Strategy for Army Security Cooperation*, Santa Monica, Calif.: RAND Corporation, MG-563-A, 2007. As of December 4, 2007:
http://www.rand.org/pubs/monographs/MG563/

Moroney, Jennifer D.P., Nancy E. Blacker, Renee Buhr, James McFadden, Cathryn Quantic Thurston, and Anny Wong. *Building Partner Capabilities for Coalition Operations*, Santa Monica, Calif.: RAND Corporation, MG-635-A, 2007. As of December 4, 2007:
http://www.rand.org/pubs/monographs/MG635/

U.S. House of Representatives, *ADM Thomas B. Fargo, U.S. Navy, Commander, U.S. Pacific Command: Testimony Before the House Armed Services Committee,* March 31, 2004. As of December 4, 2007:
http://armedservices.house.gov/comdocs/openingstatementsandpressreleases/108thcongress/04-03-31fargo.html

Thaler, David E., *Strategies to Tasks: A Framework for Linking Means and Ends,* Santa Monica, Calif.: RAND Corporation, MR-300-AF, 1993. As of December 4, 2007:
http://www.rand.org/pubs/monograph_reports/MR300/

Till, Geoffrey, *Sea Power: A Guide for the Twenty-First Century,* London: Frank Cass Publishers, 2004.